Read Aloud EPUB for iBooks
EPUB Straight to the Point Miniguide

Written and illustrated by Elizabeth Castro
http://www.elizabethcastro.com/epub
Copyright © 2011 by Elizabeth Castro. All rights reserved.

Cover image modeled in Cheetah3D, Blender, and Photoshop by Andreu Cabré
http://andreucabre.blogspot.com

Published by Cookwood Press
http://www.cookwood.com

ISBN: Print: 978-1-61150-017-2

EPUB: 978-1-61150-014-1

Read Aloud EPUB for iBooks

In June, 2011, Apple added a very cool little feature to iBooks: the ability to have an audio track that reads the book out loud. Although you've been able to add audio for some time, Read Aloud books use a special XML file to link the audio file with each individual word or sentence on a page and then allow that word or sentence to be highlighted at the same time so that the listener can follow along. It can also turn pages automatically and play a soundtrack in the background.

Read Aloud books are perfect for children's books, both for entertainment as well as to help them learn to read by following along with the text as it's read aloud. But Read Aloud books can also be great tools for language courses, song books, or anywhere else you'd like to show a transcription of an audio track, word by word.

You can only add Read Aloud capabilities to fixed layout format books, which I cover in detail in my *Fixed Layout EPUBs* miniguide. Both the Read Aloud and Fixed Layout miniguides are companion publications to my *EPUB Straight to the Point: Creating ebooks for the Apple iPad and other ereaders*. This *Read Aloud EPUB for iBooks* miniguide focuses on only those aspects of EPUB that are necessary for Read Aloud books. You may need my other publications if you are not already experienced with EPUB and fixed layout.

You'll also need a text editor with GREP capabilities. I love BBEdit, but its free sibling, Text Wrangler, will also do the trick (*http://www.barebones.com*). Of course, you can use any text editor with GREP that you like.

Creating the book

I already wrote a miniguide that explains how to create a fixed layout ebook, so I won't be repeating that information here. The basic idea, though, is that you must create a separate XHTML file for each page of your book and then position elements in specific places on the page with CSS.

Creating a template for each page

I recommend that you create a template that will serve as the basis for each page of your book. The example I'm going to use in this book is a collection of Aesop's fables, adapted into very short limerick-style rhymes by Walter Crane in 1887. Let's look at one of the first pages, that contains two fables:

Each of the images measure 1049 x 1066 pixels, and thus I've set up a template with those measurements.

 Begin the XHTML page the usual way by typing the xml declaration and the DOCTYPE.

```
<?xml version="1.0" encoding="UTF-8"?>
<!DOCTYPE html PUBLIC "-//W3C//DTD XHTML 1.1//EN"
    "http://www.w3.org/TR/xhtml11/DTD/xhtml11.dtd">
```

2 Next, create the html element, and add two special namespace declarations so that you can add play and stop buttons and a soundtrack later.

```
<html xmlns="http://www.w3.org/1999/xhtml"
    xmlns:ibooks="http://apple.com/ibooks/html-extensions"
    xmlns:epub="http://www.idpf.org/2007/ops">
```

3 Then create the head section, where you will define the viewport for your book. It's a good idea to include the viewport in the template since it must be the same for each and every page in your book. Remember that we want our pages to measure 1049 x 1066 pixels. You'll also declare the character set used for the book.

```
<head>
    <meta name="viewport"
        content="width=1049,height=1066" />
    <meta content="text/html; charset=UTF-8" />
```

4 Create the title for the book. This is required by the HTML but is not actually used by iBooks, so the importance of its content is minimal.

```
<title>Baby Aesop</title>
```

5 Next, create links to the CSS that you will use to format your book pages. I like to create a link to a set of base styles, as well as a link to the specific CSS for each page.

```
<link href="css/styles.css" type="text/css" rel="stylesheet" />
<link href="css/pagen.css" type="text/css" rel="stylesheet" />
```

When you create the individual pages of the book, you can change the *n* in *pagen.css* to correspond to the actual page number of the book, but won't have to type the whole link element from scratch.

Complete the head element and create a very basic shell for the contents of your page.

```
</head>
<body >
   <div>
   </div>
</body>
</html>
```

Save the file as *template.xhtml*, remembering to choose Unix line endings and UTF-8 character encoding. Here's what the whole thing should like once you're done.

```
<?xml version="1.0" encoding="UTF-8"?>
<!DOCTYPE html PUBLIC "-//W3C//DTD XHTML 1.1//EN"
    "http://www.w3.org/TR/xhtml11/DTD/xhtml11.dtd">
<html xmlns="http://www.w3.org/1999/xhtml"
    xmlns:ibooks="http://apple.com/ibooks/html-extensions">

<head>
    <meta name="viewport"
        content="width=1049,height=1066"></meta>
    <meta content="text/html; charset=UTF-8"></meta>
    <title>Baby Aesop</title>
    <link href="css/styles.css" type="text/css"
        rel="stylesheet" />
    <link href="css/pagen.css" type="text/css"
        rel="stylesheet" />

</head>

<body >
    <div>
    </div>
</body>

</html>
```

You can then create a copy of this document as the foundation for each page of your book.

But how do you create a fixed layout book that should be *Read Aloud*? The most important thing is that all your text should be kept out of your images, and should therefore be part of the XHTML file. Second, if you want text to be highlighted as it is read, you must mark each and every word, most often with span elements and a unique id, so that they can be associated later with the proper corresponding bit of audio. The final code might look something like this:

```
<p class="moral"><span id="W38">The</span> <span
    id="W39">grapes</span> <span id="W40">of</span>
<span id="W41">disappointment</span> <span
    id="W42">are</span> <span id="W43">always</span>
<span id="W44">sour.</span></p>
```

It's pretty ugly. While you could do this by hand, I have figured out how to get any text editor with GREP capabilities—I use BBEdit—to do it for you. This technique presupposes a few things. First, that you want to highlight each and every word when it is read in the book. And second, that each line is positioned separately on the page. At any rate, you can use this system to mark the words, and then adjust the code by hand to take into account the specific circumstances of your book.

Marking each word in the book

1 In a text editor, type all the text from a page of your book, with no formatting whatsoever. Make sure there is a return after the last line.

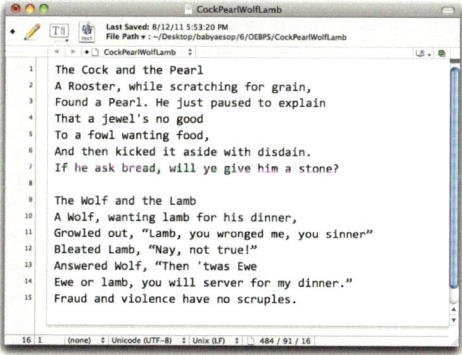

Now, before adding any other formatting, we'll use GREP to mark each and every word of text on the page.

The first step is to mark the end of each line with an *xxx* so that we can separate the lines back out at the end.

2 Choose Search > Find. Make sure the Grep box is selected. I also check the Wrap around box so I don't have to worry where the cursor is.

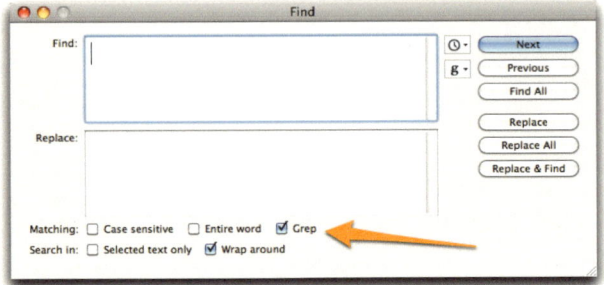

3 In the Find box, type \r to find each return. In the Replace box, type *xxx*\r. Then click Replace All.

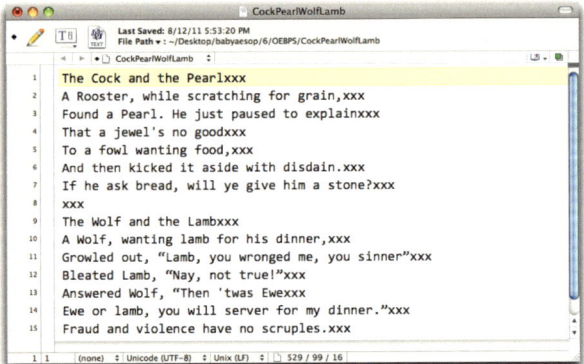

If you don't have an *xxx* at the end of the last line, it's because there was no return there. Either go back, or enter the *xxx* manually.

The next step is to put each word on a single line so that we can number them.

4 Choose Search > Find. In the Find box, type a space. In the Replace box, type \r. Click Replace All.

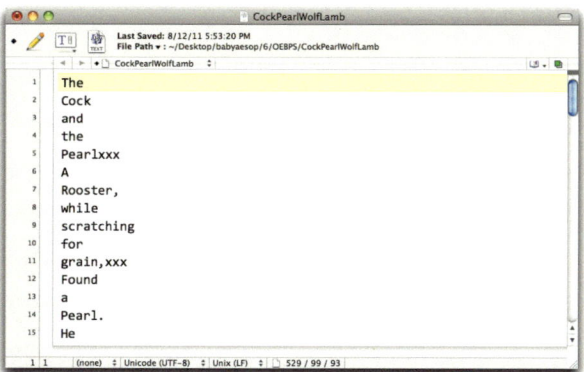

BBEdit and Text Wrangler have a automatic numbering feature that adds a number before each word. If your text editor can't number lines automatically, just do this step manually and skip to step 5. It'll still be worth it. We can then use the numbers to individually identify each word in order to map it to the audio.

5 Choose Text > Add/Remove Line Numbers. Make sure to start at 1, increment by 1, and add a space after each number.

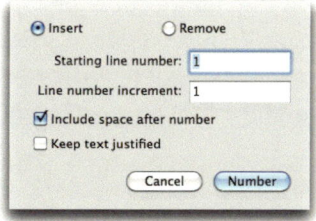

The result will look something like this. Notice that the *xxx* is still marking the division into lines:

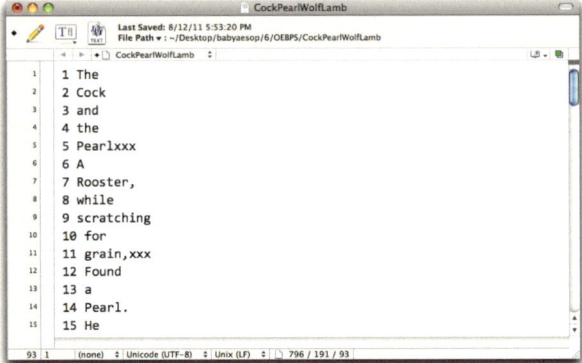

Now we want to use each line number as a unique id to identify the words so we can associate them later with the appropriate piece of audio. More GREP.

 Choose Search > Find. In the Find box, type (\d+) (.*?)(xxx)*\r. Don't forget the space after the first set of parentheses. In the Replace box, type <span id="W\1">\2</span> \3. Don't forget the space before the final *\3*. Click Replace All.

That all says, "find the number at the beginning of the line and use it with the letter *W* as a unique id in a span element around the word. If there's an *xxx*, put it outside the span element."

You may notice that the entire document is now on a single line.

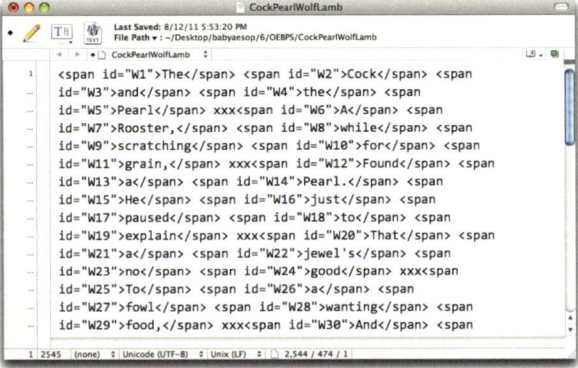

To recover our line breaks, we'll convert the *xxx's* back to returns.

Choose Search > Find. In the Find box, type a space followed by *xxx*. In the Replace box, type \r. Click Replace All.

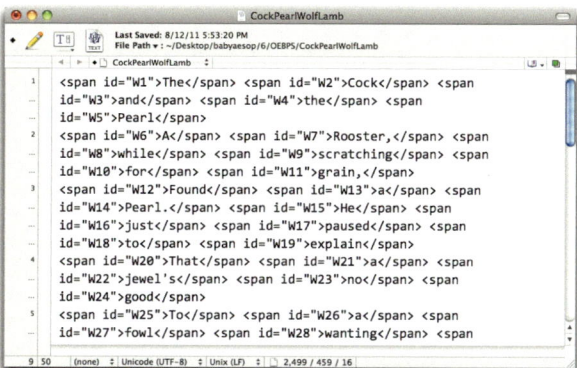

We're almost there. Now each line is on its own line. We can use GREP to add some basic HTML formatting for each line as well. Once again, we'll use the renumbering tool.

 Choose Text > Add/Remove Line Numbers, and start at 1, increment by 1, and add a space after each number.

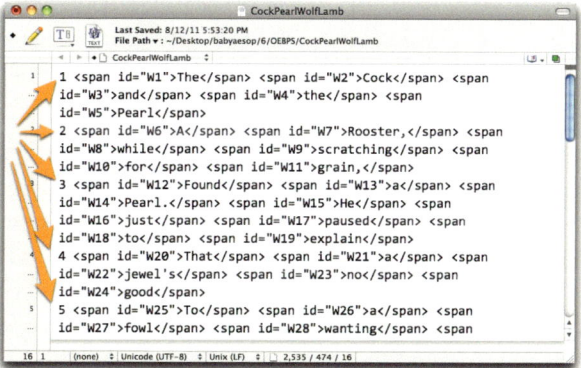

And finally, we'll convert the line numbers into classes that we can use to identify the lines and format them with CSS.

 Choose Search > Find. In the Find box, type (\d+) (.*). Don't forget the space between the two sets of parentheses In the Replace box, type <p class="line\1">\2</p>. Click Replace All.

This says, "take the number at the beginning of the line and use it in the class name when you format the content of the line with a p tag."

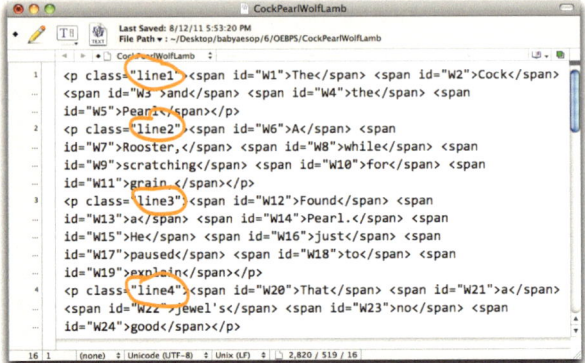

If you preview this text in a browser, you'll see that the only visible formatting is the division into lines. Nevertheless, each word is marked with a unique sequential id that we will later associate with the corresponding bit of audio from the audio file.

Once you've marked each word of text, you can add that code to your full HTML page for the book and format it as usual.

Format the pages

Once you've created the code for each page with the marked up text, you can use CSS to position the text on the page. Many people have asked how I complete this step for fixed layout books in general. Basically, I do it by hand. What I have found to be useful is to open the CSS document in a text editor, and edit the CSS while watching how the text jumps into place, adjusting it bit by bit. Apple recommends positioning text with pixel units (and not, say, ems).

1 Create @font-face rules to embed any fonts that you'll use for the text. Remember to read the licensing on your fonts to be sure you can include them in your EPUB book.

```
@font-face {
    font-family: "Trinigan FG";
    font-style: normal;
    font-weight: normal;
    src:url("../fonts/Trinigan.ttf");
    }
```

2 Set the general formatting for your paragraphs in a p element.

```
p {font-family: "Trinigan FG";
    font-size: 40px;
    letter-spacing: 2px;
    position:absolute;
}
```

3 Create a rule for the first line of text where you can specify the exact positioning it should have. Since we defined a new class for each line, you can use that class to focus the positioning on that single line.

```
p.line1 {
    top: 385px;
    left: 223px;
    font-size:34px;
    letter-spacing: 3px;
    text-transform: uppercase;
    -webkit-transform: rotate(30deg);
}
```

I keep the CSS file, the HTML file, and a preview in Safari open on my desktop so I can adjust the positioning pixel by pixel until it's just right.

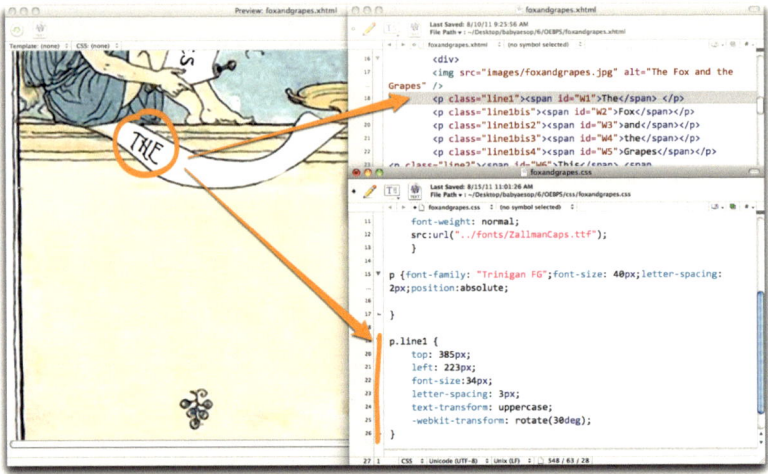

Here's the CSS for the first word. I worked it out by trial and error, adjusting the position, size, and angle values until the word fit correctly in the ribbon. Luckily, not all words have to be positioned individually!

4 Sometimes parts of a line need to be positioned separately. I create new p elements in the HTML with appended class names.

```
<p class="line1"><span id="W1">The</span> </p>
    <p class="line1bis"><span id="W2">Fox</span></p>
    <p class="line1bis2"><span id="W3">and</span></p>
    <p class="line1bis3"><span id="W4">the</span></p>
    <p class="line1bis4"><span id="W5">Grapes</span></p>
```

Then create the corresponding rules in the CSS.

```
p.line1bis {
    top: 410px;
    left: 288px;
    font-size:38px;
    letter-spacing: 3px;
    text-transform: uppercase;
    -webkit-transform: rotate(20deg);
}
```

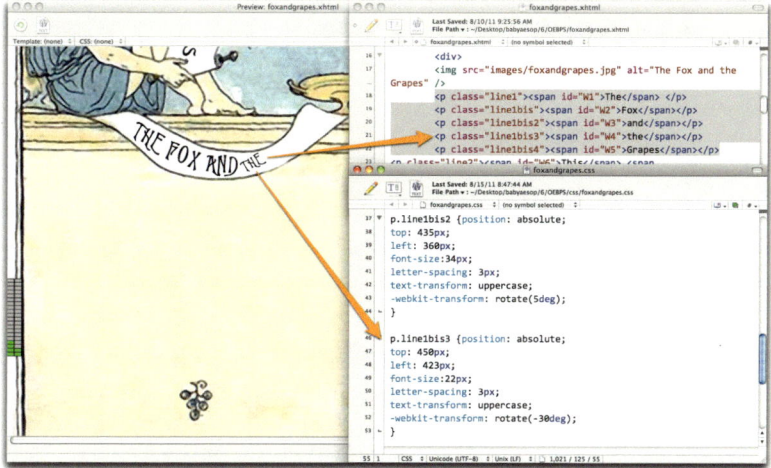

Here is what the page looks like after the CSS for the fourth word in the first line has been added. Continue in this fashion until all of the text is positioned.

 Take advantage of the first-letter CSS pseudo class to format drop-caps.

```
p.line2:first-letter {
    font-family: "Zallman Caps";
    font-size: 130px;
    line-height: 0px;
    float:left
}
```

The first real line of text has an initial drop cap, set in the font Zallman Caps and the first few words are set in uppercase. By using the first-letter pseudo-class, the text remains searchable and doesn't have to be marked up by hand.

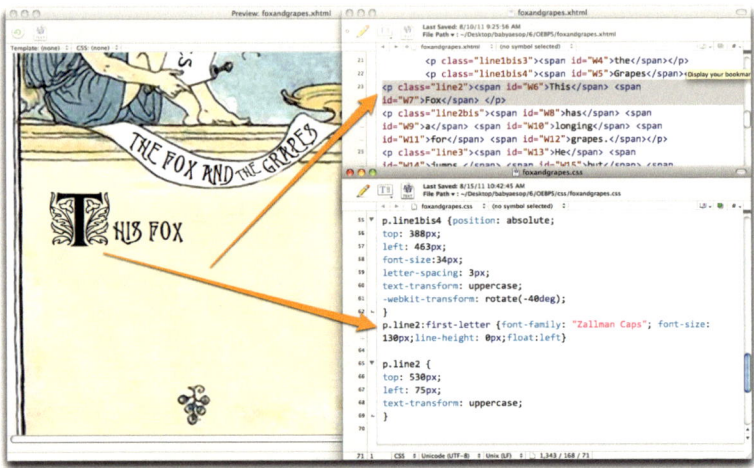

I used the first-letter pseudo class to format just the letter T *in the word* This. *Notice that no extra markup is required in the HTML.*

Formatting regular lines without extra formatting is a lot simpler:

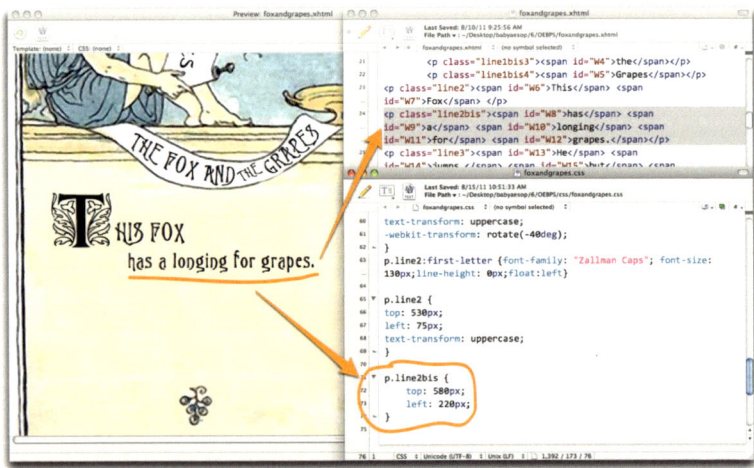

The only CSS required for a "normal" line of text is the positioning, in this case, 580 pixels from the top and 220 pixels from the left.

You can see the rest of the style sheet in the example file.

Creating the audio

The audio portion of a Read Aloud book is not an automated computer voice converting text to sound on the fly. Instead, it's an actual audio recording, and thus can contain as many voices and tracks as you want to add to your book. You can also create an ambient soundtrack, that will play in the background as someone reads your book.

One of the most important features of the Read Aloud format is the ability to highlight words as they're read, making it a good tool for kids learning to read, or even for adults learning to read in a new language. In order to associate the printed word with the audio, you must actually tag each and every word, both in the audio and in the HTML file. It's easier than it sounds.

Let's start by creating the audio file. Apple suggests using the open source program Audacity, and though I hadn't used it before, I found it a very reasonable option. If you already have an audio file, you can skip ahead to *"Marking your words in the audio file"* that describes how to tag individual words or phrases.

Apart from Audacity itself, which you can download from *http:// audacity.sourceforge.net*, you'll also need a decent microphone. I use a USB microphone that cost about $50.

Getting ready to record

There are a few adjustments you have to make to Audacity before recording.

1 Before you even open the software, be sure to plug in your microphone. Otherwise, Audacity may have trouble seeing that it's there. If you have a headset, position the microphone so that it is not too close to your mouth or noise, in order to reduce extraneous noise.

2 Launch Audacity. (Of course, you can use any audio software you like; the process should be very similar.)

3 Set the Output Volume slider all the way to the left so you don't hear your recording as you're creating it. (Otherwise, it's very distracting!)

The Output Volume slider is indicated by a speaker to the left.

4 Choose your microphone from the Input Device pop-up menu. (If it doesn't appear, make sure you launch Audacity *after* plugging it in.)

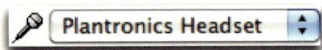

The Input Device menu has a microphone to its left.

5 Press the red record button at the top left of the window and then say a few words. When you're done, press the square orange Stop button.

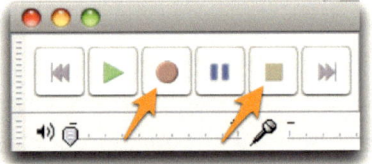

I'm not sure why they look sort of grayed out. They are *active.*

Audacity will show you a graphic representation of what you've just recorded. The taller the wave form, the louder the recording.

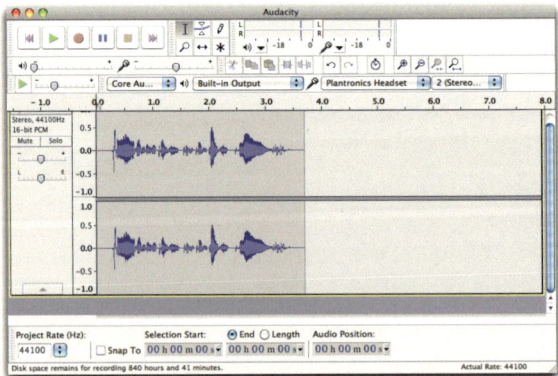

The wave forms in the center of the window represent the recorded sound.

If your recording is too low or too high, adjust the Input Volume Slider to compensate.

The Input Volume Slider

If the wave form is cut off or clipped at the top or the button, there will be distortion in the resulting audio. I've had good luck keeping the wave form to about 50-75% of the window.

You may also need to adjust the position of the microphone to reduce any ambient or distracting noises (like breathing!)

 Choose the desired output device (I use my headset) so you can listen to the audio.

The Output Device menu has a little speaker to its left.

8 Reset the Output Volume slider so you can listen to the recording. (See Step 3.)

9 Press the space bar to play the recording. You may need to place the cursor—by clicking—at the beginning of the recording. You can also choose Transport > Play.

When the test audio sounds good, you're ready to record the real thing.

Recording your book

The actual recording process is so simple it's kind of ridiculous numbering the steps.

1 Press the record button.

2 Record a single page.

3 Press the stop button.

There are, however, a number of tips to keep in mind:

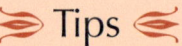

Tips

Create an individual audio file for each page. This is not offi-
cially required, but I think it makes it easier to edit the audio,
to mark individual words, and to relate the files later. It also
makes it easier to make changes to the recordings of individual pages, and lowers the pressure to make a perfect recording
of the entire book in a single try.

Read slowly and clearly.

Keep the microphone at a good distance from both your nose
and your mouth.

Be sure and save the audio recording once you're satisfied with it. Just
use the Audacity format for now. We'll talk more about exporting the
audio later.

Marking your words in the audio file

Once your audio file has been created (and saved), you're ready to
mark the individual words in the file so that you can relate them to
the words in your HTML file. This will enable iBooks to highlight the
words as the audio plays.

Marking words in the audio is a bit of a tedious process. There's really
no way around listening to the recording and completing the process
manually. However, once you start becoming familiar with the pat-
terns in the wave form, it goes pretty quickly. I wouldn't want to do a
whole novel, but a children's book is not so bad. If you are marking
whole sentences at once, you might be able to automate the process
by searching for spaces and pauses in the text. But we're going to highlight words, so I'll show you how to do it by hand.

1 Start by opening the audio file in Audacity. Use View > Zoom In
to make the wave form a little bigger and thus make it easier to
guess where the words are divided.

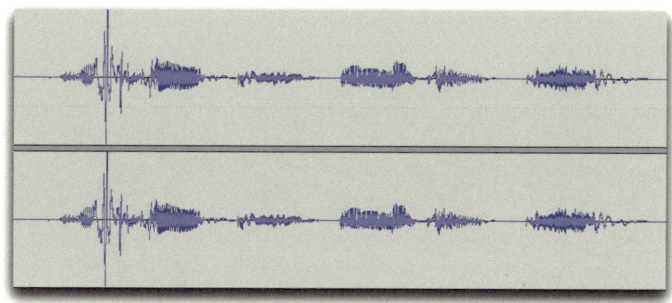

*This wave form corresponds to the words "The Fox and the
Grapes". Can you tell which pieces of the wave form correspond to
which words? Unfortunately, it's not as simple as five words and five
bumps.*

The very first thing to notice is that bigger waves denote *louder*
sounds. The places where the waves die down altogether indicate silence. Also notice that there are two sets of waves, for the
left and right speakers. The waves will be identical unless you're
using a special microphone to collect the left and right sounds
separately.

Keep in mind that less significant words—like *the*, *to*, *a*, and *an*,
in English—are practically joined to other words when spoken.
They will often be hard to separate out altogether. Don't worry
about making it too exact. Remember that the entire audio
will play at once and the reader won't notice milliseconds of
difference.

Finally, note that aspirated sounds, like the letters *t*, *s*, *k*, and *p*
will often stand out more in a wave form. You can see this in the
example at hand.

I also like to keep the original text nearby so I know just what
words I'm looking for.

2 Select what you believe to be the section of the wave form that corresponds to the first word.

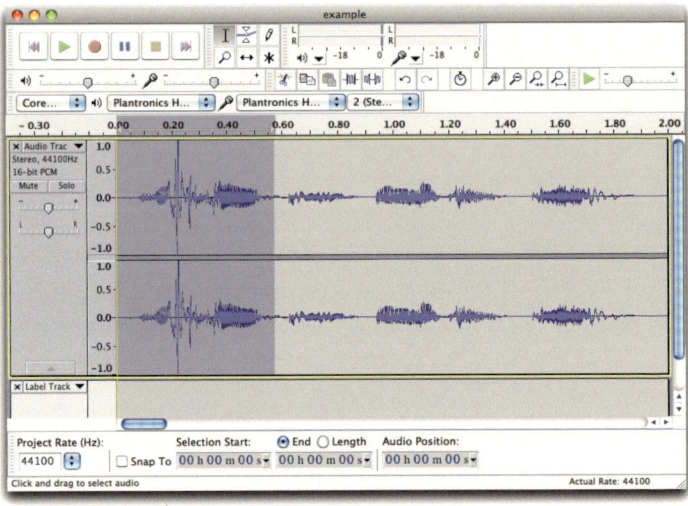

Since the very low wave form is silence, selecting the first bump is a good (but erroneous) guess.

3 Now press the space bar to listen to just that section of your audio. You'll be able to tell how accurate your selection was. In this example, the selected portion corresponds to the word *The* and the first part of the word *fox*.

4 Place the cursor over the right edge of the selection until it turns into a pointer hand. Drag the selection to adjust it so that it only covers the first word (*The*, in this case). Press the spacebar to listen again, and keep adjusting the selection until only the desired word is selected.

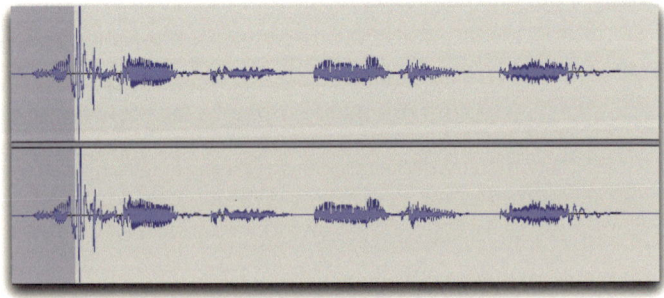

Remember that insignificant words like The *may look like they're just a tiny part of a wave form.*

5 Once you have the desired word selected in the wave form, press Command-B (or select Tracks > Add Label at Selection). Audacity creates an empty label below the audio track.

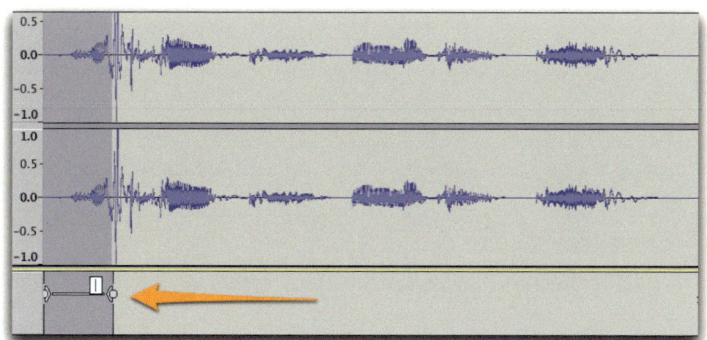

The Label track appears below the audio track in the Audacity window.

Although you can add a text label right here in the audio track, don't do it. In the next section, I'll show you how to do this step automatically.

6 Select the next word in the wave form by clicking in the wave form at the right edge of where you think the word ends and dragging toward the left. When you get to the edge of the previous word, the selection turns yellow to indicate that there is no extra sound between the two labeled pieces.

7 Again, listen to the selection (by pressing the spacebar) and adjust the right edge of the selection until it covers only the desired word.

8 When you're satisfied, press Command-B to create the next label.

Repeat steps 6-8 until all of the words in the audio track have been marked in this fashion.

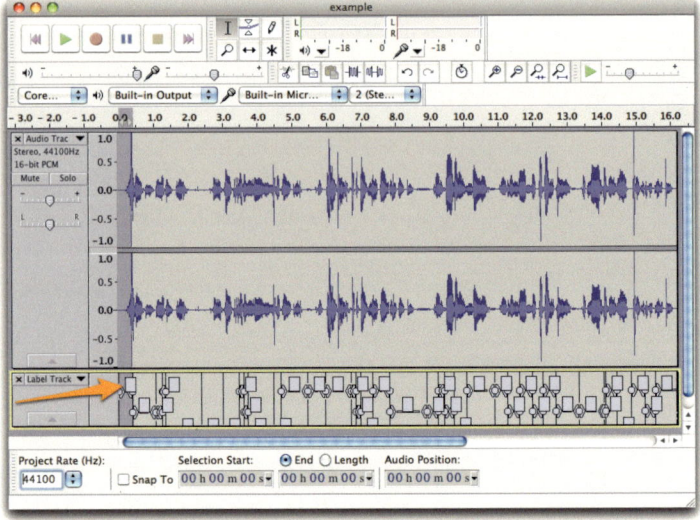

I've zoomed out quite a bit, in order to show a larger portion (but still not all) of the audio file. Notice the labels (with empty content) in the Label Track at the bottom of the window.

⇒ Tips ⇐

Don't forget to save your work periodically.

You don't have to mark every word. You can mark phrases, sentences, paragraphs, or whatever you want. Make sure that you mark the audio in the exact same way that you marked the text in your book. That way, the id numbers in the two files will match.

Use the yellow selection guide to make sure that each selection begins where the previous one ended. This will make the highlighting more fluid in your book. I think it's more important for each selection to begin where the last one left off than for them to exactly reflect a particular word.

The right and left edges of each selection are marked with a circle and a greater than or less than sign. Drag the greater than or less than sign to adjust the edges; drag the circle to move the entire selection area.

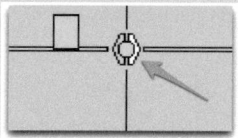

That small symbol that looks like a Darth Vader ship is actually three controls: the left selection edge, the entire selection, and the right selection edge.

Make the wave form a little bigger (Command-1 or View > Zoom in) to get a closer view of the division between words. But remember, it doesn't have to be exact. When you click on a label, you automatically go into editing mode for the text contents of the label. However, editing the labels one by one is not necessary if you use the technique I describe in the next sections. Instead, leave the labels completely empty. Press Enter to get out of label editing mode (and then you can press the spacebar to play the selection). Delete extra selections by clicking in the label and pressing backspace.

It's a good idea to leave a few seconds at the very end of each page's audio file. Otherwise, iBooks jumps to the next page too quickly (in my opinion).

If you have too much space at the beginning or end of an audio file, you can select it and press delete to get rid of it.

Keep the microphone at a good distance from both your nose and your mouth.

Audacity's online manual is very helpful and has many more details about recording than I could possibly include here. You can find it at *http://manual.audacityteam.org/man/ Main_Page*

Exporting the audio

Once your audio file is completed, you can export it into the format
that iBooks recognizes for Read Aloud books: .m4a.

Downloading and installing export libraries

The very first time you export to .m4a format, you'll have to make sure
that Audacity has the libraries required for this purpose.

1 To do so, go to Audacity > Preferences and choose Libraries in
the right hand side of the window.

2 Then click the Download button next to FFmpeg library. You'll
jump to the website that contains the latest version of the necessary files.

3 Download the libraries and use the installer to copy them to the
appropriate folder on your computer.

4 Return to the Preferences window in Audacity and click
Locate... next to the FFmpeg button. Find the libraries and then
click OK.

You're now ready to export to .m4a.

Export the audio file

Once the libraries are properly installed as described in the previous
section, exporting the file in .m4a format is quite easy.

1 Choose File > Export.... The Save As box appears in which you
can choose the folder where you wish to store the audio files.

2 Choose M4A (AAC) Files (FFmpeg) in the Format menu.

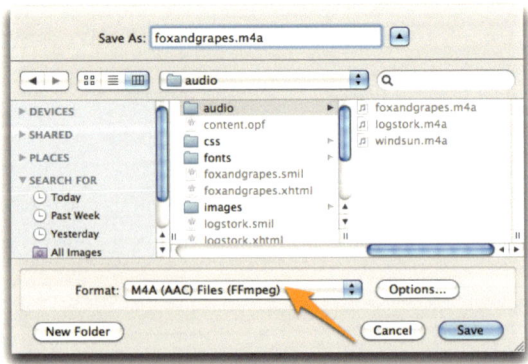

3 Choose a descriptive name without any spaces or extraneous punctuation. I use a name that matches the corresponding page of my book. Click Save.

Exporting the labels

Properly exporting the labels that indicate the starting and ending times for each word or phrase is almost as important as exporting the audio itself. Apple advises editing the text of each label individually by going to Tracks > Edit Labels, but frankly I think that is way too slow. I'll show you how to automate the whole process. Note that Audacity calls the time information about the beginning and ending of a given selection a "label" even if it doesn't have an actual text label.

1 First, make sure that you have marked each bit of audio that you want to highlight in the text as described in *"Marking your words in the audio file"*.

2 Ensure that the labels contain no text. If you have typed any text in the labels, click them to edit and then backspace to remove the text. (Don't backspace after the text is removed, or else you'll remove the label itself as well.)

3 Choose File > Export Labels…. The Save As dialog box will appear where you can choose the name and location for the saved text file that contains the starting and stopping times for each selection.

4 Click Save.

Converting the exported labels into a SMIL file

When you export labels, Audacity creates a text file that displays the beginning and ending times, in seconds, for each labeled section, separated by tabs. If you had added text to the labels, you would find it in the third column. We can use GREP to convert that file into the SMIL format that iBooks can use to associate the correct piece of the audio file to the corresponding text in the HTML file.

It is important that you have marked the audio file to the exact same degree as you marked the text. In other words, if you have marked *The Fox and the Grapes* as five separate words, you must mark them as five separate words in the audio as well. This will ensure that the numbers used to identify each word will match in the text and in the audio.

1 Open the text file containing the exported labels with a text editor that can do GREP and preferably that has an auto-numbering function.

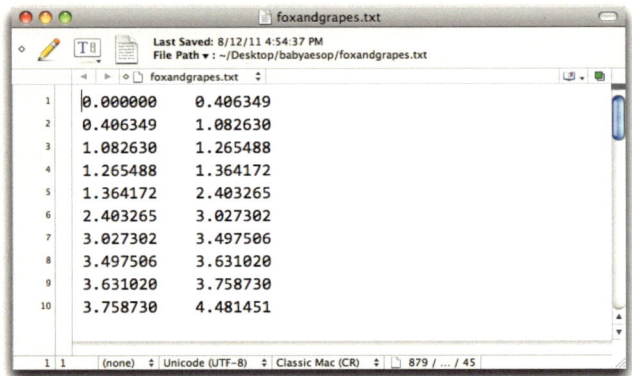

Each line of the file contains a start and stop time. The first line corresponds to the first word (or selection), the second to the second word, and so on.

2 Choose Text > Add/Remove Line Numbers. In the dialog box that appears, be sure to begin the line numbers at 1, increment each one by 1, and add a space after the number. (Just do this step by hand if your text editor doesn't do auto-numbering.)

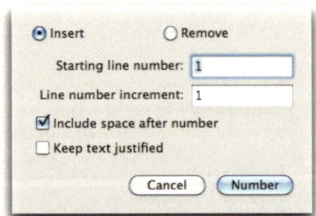

Make sure the Keep text justified box is not *checked.*

Click Number to add the numbers.

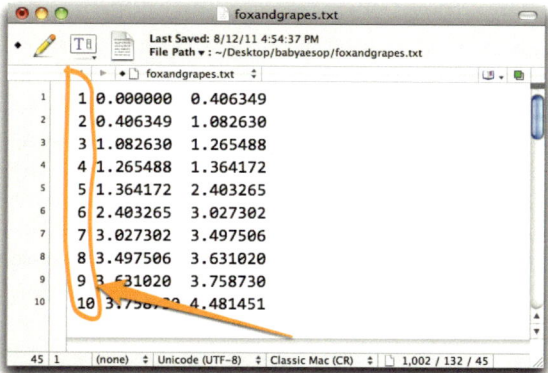

You should have the same number of lines in this document as the number of words in the XHTML document.

Now you're ready to convert each line into the format required for the SMIL file. We'll do that with GREP.

 Choose Search > Find and make sure the Grep box is checked.

In the Find box, type `^(\d*)\s(.*?)\t(.*?)\t` . This means, "find the digits at the beginning, followed by a space, followed by a bunch of text, a tab, some more text, and another tab".

In the Replace box, use `<par id="par\1">\r<text src="filename.xhtml#W\1" />\r<audio src="audio/filename.m4a" clipBegin="\2s" clipEnd="\3s" />\r</par>`. This converts what was found into par elements that tell iBooks which piece of audio is associated with which piece of text in your XHTML file. You must use the actual file names for your XHTML and audio files, respectively.

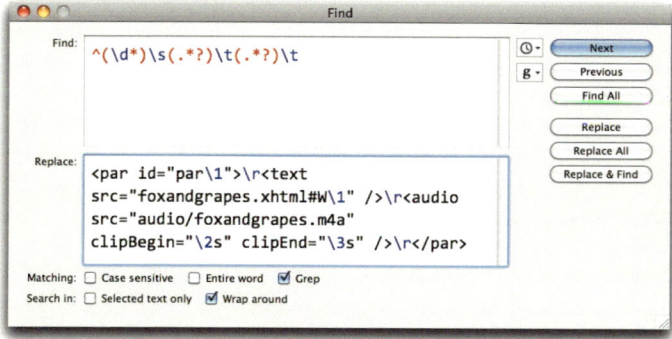

5 Click Replace All.

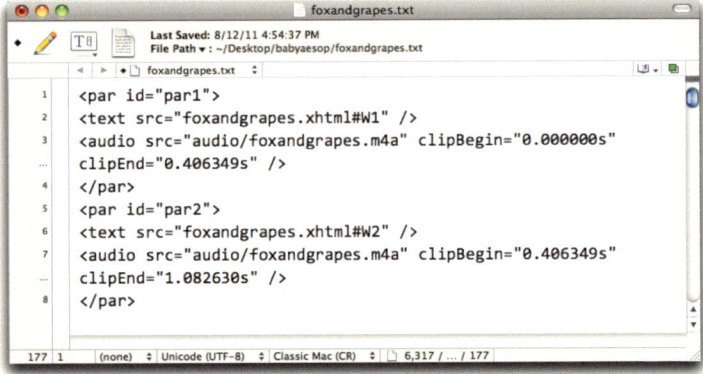

BBEdit converts each line into a proper par element in the SMIL file.

6 Insert the standard SMIL file header at the beginning of the document. You can copy the text below or use the sample file included on my website.

```
<?xml version="1.0" encoding="UTF-8"?>
<smil xmlns="http://www.w3.org/ns/SMIL" version="3.0"
profile="http://www.idpf.org/epub/30/profile/content/">
<body>
```

7 Insert the closing SMIL information at the very end of the document. Again, you can copy the text below, or use the sample file included on my website.

```
</body>
</smil>
```

Finally, save the document in Text only format with an .smil ex-
tension. I like to use the same filename as the one I've used for
the HTML and audio files so that it's easier to remember which
files go together later when I create the OPF file.

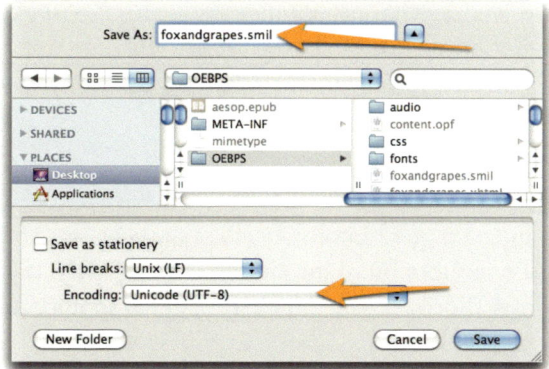

Be sure to use Unix line breaks and UTF-8 encoding.

You have to create a separate SMIL file for each XHTML file that
has associated audio.

Putting it all together

So, we have a fixed layout book whose individual words of text are marked with unique identifiers. We also have an audio file of those words. Finally, we have a third file that associates the marked words with the corresponding portion of the audio. The only thing left is to tell iBooks which files go together. To do that, we edit   the content.opf file.

Remember that the content.opf file (described in complete detail in my book *EPUB Straight to the Point*) lists the files that are used in a given EPUB book, and also gives information about what sort of files these are. We'll add one more bit of important information to the content.opf file which allows iBooks to see which audio files should be used with which pages of your book.

Listing the files in the manifest

The first step is to list each file used in the EPUB. In our example, we will have one XHTML file, one SMIL file, and one audio file for each page.

1 In the manifest section of the content.opf file, declare the XHTML file as usual.

```
<item id="p2" href="foxandgrapes.xhtml"
    media-type="application/xhtml+xml" />
```

2 On the next line, add the item information for the SMIL file, as follows. Remember to use a unique value for the id element. I use one that is (relatively) easy to type but still has something to do with the corresponding XHTML file.

```
<item id="foxandgrapessmil" href="foxandgrapes.smil"
    media-type="application/smil+xml"/>
```

3 Add the item information for the audio file. Again, the id ele-ment must have a unique value.

```
<item id="foxandgrapesaudio" href="audio/foxandgrapes.m4a"
    media-type="audio/mpeg"/>
```

4 Finally, in order to associate the XHTML file with the SMIL file, which will then call the audio file, you must add a media-overlay attribute to the XHTML item element that you created in step 1, as follows. The value of the media-overlay attribute corresponds to the value of the id element of the SMIL item

```
<item id="p2" href="foxandgrapes.xhtml"
    media-type="application/xhtml+xml"
    media-overlay="foxandgrapessmil" />
```

The final declarations for the XHTML, SMIL, and audio for each page of your book will look something like this:

```
<manifest>
...
<item id="p2" href="foxandgrapes.xhtml"
    media-type="application/xhtml+xml"
    media-overlay="foxandgrapessmil" />
<item id="foxandgrapessmil" href="foxandgrapes.smil"
    media-type="application/smil+xml"/>
<item id="foxandgrapesaudio" href="audio/foxandgrapes.m4a"
    media-type="audio/mpeg"/>
...
</manifest>
```

5 Repeat these steps for each page in your book that has associated audio.

Styling the audio highlight

If you want the audio to be highlighted when the book is read, you'll have to define the desired styles in the CSS file. To do, style the special pseudo-class opub-media-overlay-active which is automatically (and invisibly) applied to elements as their corresponding audio is played.

1 Open the CSS document for your book.

2 Create the selector for all elements with the class -epub-media-overlay-active as follows by typing a period followed by that class name. Don't forget all the dashes, particularly the initial one and don't add any spaces.

```
.-epub-media-overlay-active
```

3   Define the styling that should be applied when the word or selection is read.

```
.-epub-media-overlay-active {
    color:red;
    }
```

Testing the book

Now that you've got the whole book set up, you'll want to test it in an iPad, iPhone, or iPod touch to make sure it works as expected. You can find information about compiling an EPUB file and transferring it to your iOS device in my *EPUB Straight to the Point* book.

1   Open up the book and click anywhere in the book to make the navigation system appear. If you've followed the steps in this miniguide correctly, you'll see a speaker in the upper-right corner of the screen.

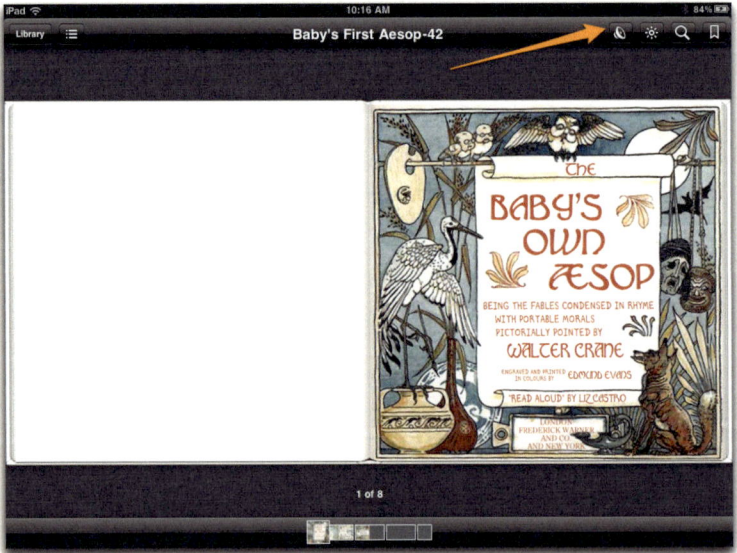

2 Press the Read Aloud speaker to view the Read Aloud menu options.

You can adjust the volume, have iBooks turn pages automatically, and then click Start Reading to begin the book.

The Read Aloud menu is pretty self-explanatory. Note that if you choose Automatically for Turn Pages, iBooks will read each page and then jump to the next page until it gets to the end of the book. If you choose Manually, iBooks will read a two-page spread and then curl up the bottom-right corner of the page to indicate that the audio portion is complete. The user can then click the curled up corner to continue with the book. iBooks automatically begins reading on each new page.

When the audio on a spread is complete, the reader can choose to continue at their own pace by clicking the curled up page corner.

Adding extra features

There are a couple of extra features that you might be interested in adding to your Read Aloud books, including a background soundtrack and buttons that can control the playback on a given page.

Adding a background soundtrack

Apple's iBooks app also lets you play music in the background as someone reads the book. You can choose a single piece of music for your entire book and iBooks will use it seamlessly from one page to the next. You can also choose a separate audio file for each two-page spread.

1 Make sure you have included the EPUB namespace declaration in your html elements as described earlier.

```
<html xmlns="http://www.w3.org/1999/xhtml"
    xmlns:ibooks="http://apple.com/ibooks/html-extensions"
    xmlns:epub="http://www.idpf.org/2007/ops">
```

2 Add an audio file to your book pages in the usual way.

```
<audio src="audio/JosephHaydn_06.m4a" />
```

3 Within the audio element add a special EPUB attribute to set it up as a background soundtrack.

```
<audio epub:type="ibooks:soundtrack"
    src="audio/JosephHaydn_06.m4a"></audio>
```

4 Theoretically, if you don't explicitly call for controls, the audio element will be invisible. Apple suggests you style the audio element right off the page, just in case, for example, with this CSS. I don't really think it's necessary.

```
audio {
    position: absolute;
    top: -30px;
}
```

⇒ Tips ⇐

Background audio will not play unless your reader has chosen the On position next to Soundtrack in the Read Aloud menu.

The Soundtrack option will only appear in the Read Aloud menu for those spreads that contain a page for which you've added a soundtrack.

iBooks controls soundtracks for each two-page spread. So, if you've set a soundtrack for one page in a two-page spread, the menu option will be available and the soundtrack will play for both pages. Even if you choose a different soundtrack for each page, iBooks will continue to play the left page's soundtrack for the right page and will only play a new soundtrack when you turn to a new spread.

Be careful that the audio for the background soundtrack doesn't drown out the narration of your book. You may need to adjust the levels in a program like Audacity since iBooks doesn't let the reader control the volume of the soundtrack and the narration separately.

Each audio file will play over and over again in a loop from one page to the next, or on a single spread if the reader happens to pause there.

My favorite source for royalty-free music for background tracks is *http://MusOpen.org*. (Many thanks to all those virtuosos who share their music with the rest of us.)

Creating buttons for each page

You can create individual stop and start buttons on each page of your book if you like. This gives the reader some control over beginning to read a given page. Otherwise, the reading always starts on the left page of a spread. Nevertheless, the stop button is not a pause. Starting is always from the beginning.

You can create buttons all day in an EPUB book but to get the button to control the read aloud feature, you have to add a special iBooks-specific attribute. There are three possible values: start, stop, and start-stop. The first two are pretty self-explanatory, the third starts or stops the reading, depending on what's it doing at that moment.

The trickiest part of creating start and stop buttons is figuring out what they should look like, and particularly what they should look like when they are active. That is, if you create a start button and someone clicks it, what should happen? People are used to buttons reacting. Luckily Apple offers code that you can use to change the aspect of a button once it is clicked.

Note that buttons work differently depending on how the book is displayed. If a user is viewing a two-page spread, the buttons apply to the entire spread. If a user has zoomed in to an individual page, the buttons apply only to that page.

Creating a startstop button

Let's start with the startstop button. A startstop button either starts the playback if it was stopped or stops it if it was playing. Since it is a single button, the same thing will show both when the audio is playing and when it is stopped. You can use color to show the active state.

1   If you haven't done so already, add the special iBooks namespace declaration to your HTML document for each page that will have a startstop button.

```
<html xmlns="http://www.w3.org/1999/xhtml"
    xmlns:ibooks="http://apple.com/ibooks/html-extensions"
    xmlns:epub="http://www.idpf.org/2007/ops">
```

2   Next, create a startstop button by creating a p element with the special ibooks:readloud attribute set to *startstop*.

```
<p ibooks:readaloud="startstop"
```

3   Give the button a unique id. In this example, we'll use *rass* (read aloud start stop) and don't forget to close the initial p element with a greater-than sign.

```
<p ibooks:readaloud="startstop" id="rass">
```

Next, we'll enter the text content of the startstop button. You can use words, but since you won't be able to change the text according to its state (nor translate it for speakers of other languages), I recommend using symbols.

4 Type `&#x2735;` which is a symbol for a star ✳. Of course you can use any symbol (or text) you want.

```
<p ibooks:readaloud="startstop" id="rass">&#x2735;
```

5 Finally, type the closing p element.

```
<p ibooks:readaloud="startstop" id="rass">&#x2735;</p>
```

⪦ Tip ⪧

You can download my free guide to *Using Zapf Dingbats in EPUB* from the Apple iBookstore: *http://itunes.apple.com/us/book/isbn9781611500219*. It reminds you how to code Zapf Dingbats in an EPUB file and gives a handy table that shows you the available symbols.

Styling a startstop button

Creating a startstop button is only half the battle. You'll also want to style the button so that it reacts when a user presses it.

1 In the CSS file for the page, type html followed by a space and then p followed by a number sign (#) and the id for your startstop button. This is the selector that corresponds to your startstop button.

```
html p#rass
```

2 Define the rules to format the startstop button when the audio is *not* playing. In this case, I have positioned it absolutely, colored it green and then given it a white background and a rounded corner border, so that it looks like a button.

```
html p#rass {
   top: 955px;
   left: 800px;
   padding: 2px 6px 4px 8px;
   color: green;
   background: white;
   line-height: 40px;
   border: 2px solid black;
   border-radius: 15px;
}
```

If the audio is not playing, the button is green.

3 To set the rules for how the startstop button should be styled when the audio *is* playing, create a new selector. Type html.-ibooks-media-overlay-enabled followed by a space, a p, a number sign (#) and the id of your startstop button.

```
html.-ibooks-media-overlay-enabled p#rass
```

4 Write the style rules to determine how the button should change when the audio is playing. In this case, I just want it to go from green to red, which will remind the reader that it can now be used to stop the audio.

```
html.-ibooks-media-overlay-enabled p#rass {
    color:red;
}
```

If the audio is playing, the button turns red, signifying that it is now a stop button.

Creating separate stop and start buttons

The biggest disadvantage of a startstop button is that the content is the same whether the audio is playing or is stopped. If you want different content to appear, use individual start and stop buttons.

1 If you haven't done so already, add the special iBooks namespace declaration to your HTML document for each page that will have start and stop buttons.

```
<html xmlns="http://www.w3.org/1999/xhtml"
    xmlns:ibooks="http://apple.com/ibooks/html-extensions"
    xmlns:epub="http://www.idpf.org/2007/ops">
```

2 Next, create a start button by creating a p element with the special ibooks:readloud attribute set to *start*.

```
<p ibooks:readaloud="start"
```

3 Give the button a unique id. In this example, we'll use *raplay* (read aloud play). Don't forget to close the initial p element with a greater-than sign.

```
<p ibooks:readaloud="start" id="raplay">
```

Next, we'll enter the text content of the start button. You can use text, like *play*, or if you want it to be more universal, use symbols.

4 Type ➤ which is a symbol for a right pointing arrow, like a play button. Of course you can use any symbol (or text) you want.

```
<p ibooks:readaloud="start" id="rass">&#x27A4;
```

5 Finally, type the closing p element.

```
<p ibooks:readaloud="start" id="rass">&#x27A4;</p>
```

6 Follow the same steps to create the stop button, but use *stop* for the value of the ibooks:readaloud attribute, and use a unique id. I use ✖ (a stylized x) for the stop symbol.

```
<p ibooks:readaloud="stop" id="rastop">&#x2716;</p>
```

Styling the start and stop buttons

I like to style the start and stop buttons right on top of each other and then use the display property to show only one or the other, depending on whether the audio is playing or not.

 To target the buttons, type html followed by a space, a p, number sign (#), and the id of the start or stop button.

    html p#raplay

    html p#rastop

2   Define the rules, within curly brackets as usual, to format the buttons for when the audio is *not* playing. In addition to the absolute positioning in the same location, I have made the play button green and the stop button red, given them both white backgrounds with rounded borders, and hidden the initial display of the stop button.

    html p#raplay {
        top: 955px;
        left: 700px;
        color: green;
        padding: 0px 6px 0 10px;
        background: white;
        border: 2px solid black;
        border-radius: 15px;
    }

    html p#rastop {
        top: 955px;
        left: 700px;
        padding: 0px 6px 0 10px;
        color: red;
        display: none;
        background: white;
        border: 2px solid black;
        border-radius: 15px;
    }

When the audio is not playing, the green arrow button is displayed.

3 In order to define how the buttons change when the audio *is* playing, we'll use the same html.-ibooks-media-overlay-enabled selector as before, but now with the id of our start and stop buttons.

html.-ibooks-media-overlay-enabled p#raplay

html.-ibooks-media-overlay-enabled p#rastop

4 Then when the audio is playing we'll make the play button disappear (with display:none) and the stop button appear (with display:block).

```
html.-ibooks-media-overlay-enabled p#raplay {
    display: none;
    }
html.-ibooks-media-overlay-enabled p#rastop {
    display: block;
}
```

*When the audio is playing (notice how Grapes is highlighted), the
stop button is displayed with its stylized red x.*

ꙮ Tip ꙮ

There seems to be a bug with buttons on the last spread.
Instead of starting the audio playback, they jump the reader
to the first page of the book.

Samples and more information

The plan is to make the full Read Aloud version of *Baby's Own Aesop* available for purchase on my website and/or from the Apple iBookstore. Meanwhile, the sample file used as an example in this book can be downloaded from *http://www.elizabethcastro.com/epub/samples/ BabysOwnAesopSample.epub*

You can find more information on creating EPUB files in my book, *EPUB Straight to the Point: Creating ebooks for the Apple iPad and other ereaders*, which is available from my website—http://www.elizabethcastro.com/epub—in EPUB format, and from Peachpit, Amazon, Barnes & Noble, and other booksellers in print.

If you need more information about creating Fixed Layout EPUB, you might consider my *Fixed Layout EPUB* miniguide, available on my website.

I also have published a number of articles about creating EPUB files on my curiously named blog, Pigs, Gourds, and Wikis (*http://www. pigsgourdsandwikis.com*) and tend to tweet a lot of EPUB related material through my Twitter account: *@lizcastro*

I'm always interested in comments and suggestions. You can post comments to my blog, send me messages on Twitter, or use my email form *(http://www.elizabethcastro.com/epub/contact.html)*.

Index

Made in the USA
Monee, IL
15 April 2026